P9-AFY-683

THE LIBRARY OF
FUTURE ENERGY

BIOFUEL POWER OF THE FUTURE

NEW WAYS OF TURNING ORGANIC MATTER INTO ENERGY

CHRIS HAYHURST

THE ROSEN PUBLISHING GROUP, INC.
NEW YORK

Published in 2003 by The Rosen Publishing Group, Inc.
29 East 21st Street, New York, NY 10010

First Edition

Library of Congress Cataloging-in-Publication Data

Hayhurst, Chris.
Biofuel power of the future : new ways of turning organic matter into energy/ by Chris Hayhurst.
p. cm. — (The library of future energy)
Summary: Examines the pros and cons of using plant and animal waste to meet our growing demand for electricity.
Includes bibliographical references and index.
ISBN 0-8239-3659-7 (lib. bdg.)
1. Biomass energy—Juvenile literature. [1. Biomass energy.] I. Title.
II. Series.
TP339 .H39 2002
333.95'39—dc21

2002006500

Manufactured in the United States of America

CONTENTS

	Introduction	5
1	The Basics	9
2	History and Development	25
3	The Politics of Energy	33
4	The Pros and Cons of Bioenergy	39
5	Into the Future	49
	Glossary	56
	For More Information	58
	For Further Reading	60
	Bibliography	61
	Index	62

INTRODUCTION

Have you ever considered how things work? How, for instance, does your computer turn on with the push of a button? And how do phones, fax machines, lights, refrigerators, and televisions ring, beep, light up, hum, and get perfect reception? How do cars move? And, come to think of it, how can planes take off and fly across the sky?

The answers, of course, are complex. But if you were to boil them down to the nitty-gritty basics, you'd need just one word: energy. As defined in the dictionary, energy is the capacity for doing work. In other words, when something has energy, it's able to work.

Where does energy come from? That, too, is a tricky question. Scientists, politicians, environmentalists, builders, and others have wrestled with this question for years. Everyone agrees that energy comes in many forms, from many places. Plants get energy directly from the Sun. People get energy by eating food. Other forms of energy, like the electricity necessary to run a microwave or the fuel required to operate an engine, are generated from Earth's natural resources. Examples of energy-supplying resources include things like coal, oil, wind, water, and the subject of this book, biomass. Biomass includes things like plants, wood, and animal waste.

People often disagree about which energy-supplying natural resources humans should tap for their many energy needs. Some people feel that fossil fuels, like coal and oil, can and will supply the world with abundant energy for years to come. Other people, pointing to evidence that shows coal and oil supplies are limited, believe we should rely more on "renewable" energy resources, like the wind or solar energy. Renewable energy resources constantly replenish themselves. They can never be depleted.

There are arguments for and against other sources of energy, too. Some have to do with money and economics. Others concern the environment. Still others are political, as local communities, states, nations, and continents strive to secure enough energy to meet their needs.

Biofuel is generated from a wide range of raw organic materials, some of which may appear to the untrained observer to be odd energy sources. The image on the left shows a lagoon of cow manure from which methane gas is converted into electricity. In the image on the right, an environmental scientist displays a bottle of envirodiesel fuel made from vegetable oil.

The fact is, we all need energy to survive. You can feel this every time you eat. If you stopped eating, it wouldn't be long before you "ran out of energy." But just as we need regular meals to keep us going, the machinery and technology we rely upon in our day-to-day lives also need energy to keep them going.

Energy is critical to society. As the world's population grows and our energy needs expand, the importance will magnify. Where will we get the energy of the future? Perhaps from plants we grow or garbage we recycle!

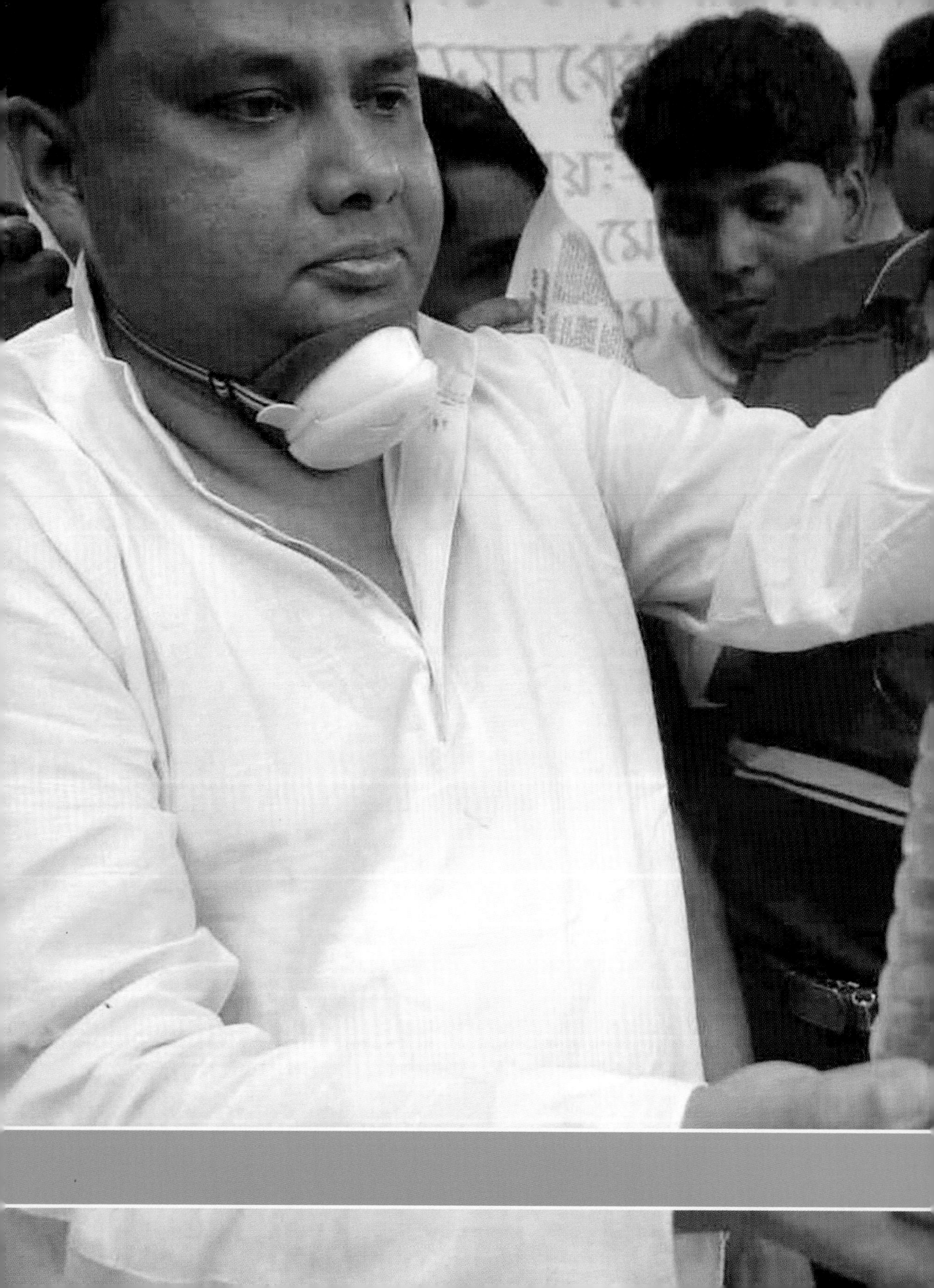

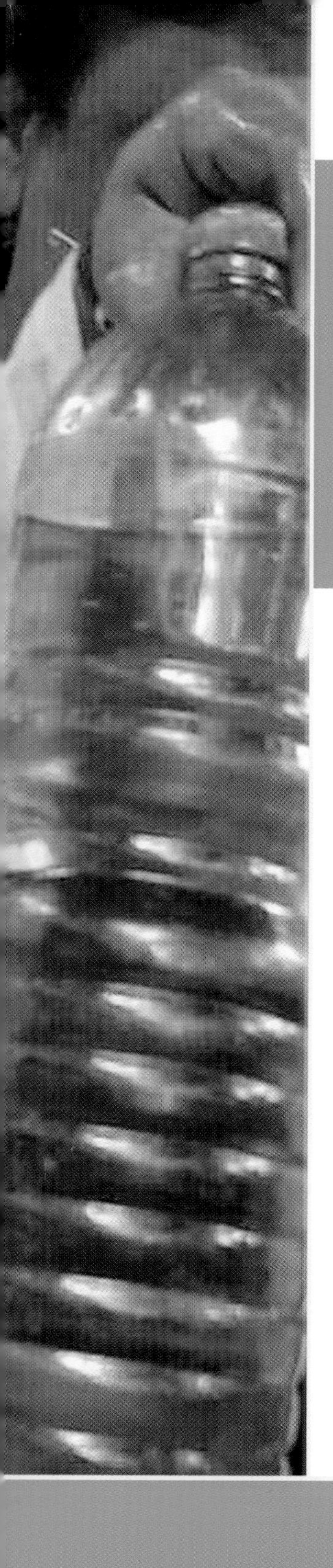

1 THE BASICS

Many experts are convinced that the wave of the future is in renewable energy. Renewable energy is energy that never runs out. It is replaced, or "renewed," as it's used.

There are many sources of renewable energy. Two types of renewable energy that are becoming increasingly popular are solar energy and wind energy. Solar energy, or solar power, is energy taken directly from the Sun. You may know someone who lives in a house with high-tech solar panels on the roof that capture sunlight and convert it to electricity. Wind power is common in parts of the the world where winds are strong and consistent. Windmills generate power when the wind is powerful enough to turn their propeller-like blades.

AN INTRODUCTION TO BIOMASS

Like the Sun and wind, biomass is another source of renewable energy. Biomass is organic material such as plants and trees. Agricultural waste like cow and chicken manure is also biomass.

Biomass can be found all over the world, wherever there are plants or animals. Seaweed is biomass. So is human sewage. Biomass is renewable because, as in the case of trees and plants, it can grow back or is naturally replaced after it is harvested. Fossil fuels, once they have been dug up and mined from the earth, take millions of years to replace themselves. Biomass, however, can be replenished almost immediately. Sometimes this process is natural, like when grass grows back after it's mowed. Other times, as is the case with many trees, someone must plant more.

A LINK TO THE SUN

There are many ways to obtain energy from biomass. One way has to do with the Sun, that great big ball of fire in the sky. The Sun is the primary source of energy for every living organism on Earth. Energy from the Sun, in the form of heat and light, travels through the solar system, ninety-three million miles to Earth. You can feel heat energy every time you go outside on a sunny day. Some animals, especially reptiles such as lizards and snakes, rely on heat energy to control their body temperatures.

You can see light energy in colors, as in the spectrum of a rainbow. As light waves reflect off moisture in the atmosphere, the colors of the rainbow appear.

When it comes to biomass, and, for that matter, all life on Earth, the Sun's light energy is what is important. Light travels through space in tiny packets of energy called photons. You can see light energy in colors. Colors are nothing more than reflected light. Whichever color you see depends on how much energy the photons reflect.

PHOTOSYNTHESIS

Plants are the only organisms that are able to capture light energy directly from the Sun. Through a unique process called photosynthesis, plants take light energy and convert it to food. Photons of light

A campfire is a common—though inefficient and unhealthy—form of bioenergy.

energy enter a plant through its leaves. Then, in a complex sequence of chemical events involving water and a natural gas called carbon dioxide, the photons convert into energy-packed food. Some of this food is used by the plant. The rest is stored inside the plant as special microscopic compounds called carbohydrates. When we turn to biomass as a source of energy, it's the carbohydrates that we're after.

GETTING TO THE ENERGY

Biomass energy can be extracted from plant matter by setting it ablaze. When our ancient ancestors discovered fire, they also discovered a form of energy. As they burned wood, they burned biomass. Burning biomass is one of the ways to get to the energy inside it.

Burning biomass produces heat. Cozy up to a campfire on a chilly night and feel the heat. People around the world burn biomass for heat all the time. They burn wood in stoves and fireplaces to warm their homes. They light fires to cook their food.

Unfortunately, open wood fires are terribly inefficient and often unhealthy. Much of the heat is wasted as it escapes into the air. And smoke, of course, is the last thing you'd ever want to breathe. Many people develop respiratory problems from breathing uncontrolled smoke. Smoke from wood fires is a major air pollutant in many parts of the world.

While an open fire shows us one way that we can obtain biomass, there are plenty of other ways, too. In fact, when it comes to producing bioenergy on a large scale, the process is far more scientific. The biomass is still burned, but the burning takes place at a power plant where the heat is carefully captured. The heat is used to evaporate water into steam (like the steam you see when you take a hot shower). The steam, in turn, can be funneled at high pressure to turbine engines. The pressure from the steam pushes against the turbine blades and causes them to rotate. This rotational movement powers a generator to make electricity.

Another way to get energy from biomass is to convert it into biofuels. Biofuels include liquids like ethanol and biodiesel that can be used to power vehicles. Some vehicles can run on 100 percent biofuel. Others run on a mixture of biofuel and regular or diesel gasoline.

Ethanol is made by converting the carbohydrates found in plants into sugars, then aging the sugars until they turn into alcohol. The conversion process is called fermentation. It is similar to the process

Corn farmers like Glenn and Sharon Arfstrom of Willmar, Minnesota, have found a new market for their corn among ethanol producers. Their van is designed to operate on fuel ranging from regular gasoline to a fuel blend that is 85 percent ethanol.

of brewing beer. According to the U.S. Department of Energy, about 1.5 billion gallons of ethanol are made every year in the United States. Ethanol can be used as fuel for gas-powered vehicles. It's commonly used as an additive in gasoline, but it can also be used on its own.

Biodiesel, which is not an alcohol, is composed of substances called fatty acid alkyl esters, which contain natural carbon. People can make biodiesel from the oils and fats of plants or animals. Using a process called transesterification, the oils or fats are combined with an alcohol, such as ethanol, to make "fatty esters" like ethyl or methyl ester. These esters can be mixed with regular diesel fuel and used to run diesel vehicles. Like ethanol, biodiesel can also be used by itself. Biodiesel is far less common than ethanol. The Department of Energy reports that just five million gallons of biodiesel were produced in the United States in 2000.

The owner of a biodiesel Volkswagen Jetta displays a vial of vegetable oil that he mixes in an organic fuel to run his car. The car travels about 40 miles per gallon using the biofuel—10 miles per gallon farther than with gasoline.

American biodiesel producers most often turn to recycled cooking oils or soybean oil for their biomass. "Neat" biodiesel, which is 100 percent biodiesel, as well as biodiesel blended with regular diesel, is commonly used in boat engines, in public buses, and in government vehicles. The next time you ride a local bus, ask the driver about the fuel. You may be surprised by the answer!

Biomass Sources

Biomass exists in many shapes, forms, and sizes, in every nook and cranny of the world. But a lot of the world's biomass is off-limits when it comes to producing energy. Redwood trees, for example, could be used as biomass. But no one would suggest chopping down these ancient giants to burn for electricity.

Many people feel that the earth is full of endangered jungles and forests that should not be used for energy. They feel these areas are better left to support wildlife. Since the world's entire food supply would vanish without plants, these people are concerned that we don't use every last bit of vegetation to make energy.

Still, the biomass supply is basically limitless—if you know where to look. A few of the most promising sources of energy-producing biomass include forests, agricultural wastes, and waste from landfills. Some farmers grow crops specifically to supply energy. Here's a rundown:

The Forest Industry

One of the best places to find biomass is among the "waste products" of the forest industry. According to the United States Department of Energy, every year 100 to 280 million tons of forest material—things like wood scraps, sawdust, twigs, and branches—go unused. Often the only thing logging companies want from the trees they cut down are the long, straight trunks. They can hew the trunks into logs and boards, which can then be used to build houses, tables, shelving, and other wood products. Huge piles of otherwise useless wood scraps are common at lumber mills, as is sawdust from the milling process.

Wood waste also comes from commercial tree farms, where trees are grown specifically for lumber, for sale as Christmas trees, and for sale to private landowners who wish to plant them on their property.

Wood for biofuel can come from large forests where undergrowth has accumulated to dangerous levels. In the past, many forest fires were entirely natural. Fires regularly roared through the forests, burning away small trees and brush and scorching trunks. But the fires did not always destroy the larger trees that made up the bulk of the forest. As difficult as it is to believe, these fires maintained the natural balance of the forests. They prevented forests from becoming too thick with vegetation and kept the competition for sunlight, soil nutrients, and water at healthy levels.

In the last half century, however, natural forest fires have been prevented for many reasons, such as to protect houses and preserve popular hiking and camping areas. This has led to such overcrowding that some forests are not as healthy as they once were. In addition, the presence of all this extra vegetation increases the risk of a fire that could destroy the entire forest, not just the undergrowth.

Some forestry experts argue that the solution to this dilemma is to thin the crowded forests by carefully logging and harvesting

A tractor-trailer unloads wood chips at Woodland Biomass Power Ltd. in Woodlawn, California. The company serves about 25,000 homes with the electricity it generates from the wood chips.

the undergrowth. The wood and other biomass could be used for things like bioenergy. Others worry that logging companies will use the "thinning" argument to justify logging where it isn't needed as an excuse to cut down old-growth forests. This fear has led some to say that natural forests should be entirely off-limits when it comes to potential biomass sources.

Agricultural Waste

A less controversial source of biomass is the leftovers from the agricultural industry. If you've ever planted a garden, you know just how much extra "stuff" comes up with vegetables. We normally consume just the fruit of a plant, not the tough, fibrous stalks and stems, or leaves. We can compost a good portion of the inedible plant parts, turning nutrients back into the soil for the next season's harvest. But much of the plants would be better used for bioenergy.

Inedible plant parts that farmers would have thrown away or used in compost decades ago are now being recycled to produce energy.

The home garden is, obviously, very small-scale. Unless you live on a giant farm, it's highly unlikely that you could pull all the energy you need to go about your daily life from your personal plot. But professional farmers—those working hundreds or thousands of acres at a time and harvesting tons of crops each season—are another story. According to the U.S. Department of Energy, farmers in the United States generate more than ninety-five million tons of agricultural "waste" every year.

When they harvest crops like corn, for instance, farmers often leave much of the plant in the field. The parts that are left behind include corn leaves, stalks, and cobs, which are full of energy-rich carbohydrates. These leftovers, called stover, can be turned into bioenergy.

Another, less glorious form of agricultural waste is manure. Because animals eat plants, their manure holds a large amount of energy. It can be collected and used as a source of fuel in power plants. Some farms, especially large cattle, poultry, and hog farms, have what are called manure lagoons on site. These giant pools of animal sewage are major air polluters because they are a source of methane gas. Methane gas is a carbon-containing compound, and it, too, can be collected and burned for energy.

Landfill Waste

Still another source of biomass is the billions of tons of garbage that ends up in landfills every year. In fact, so much trash is making its way into the nation's landfills that the landfills are getting full. Some landfills have been closed down. Others are nearing capacity and are charging higher fees as a result.

A lot of this garbage is truly useless junk—things like rotting bed box springs, broken plastic toys, and other nonrecyclable trash from homes and offices. But a huge part of it is vegetation. Every weekend, millions of people make their way to their local landfills hauling countless tons of plant matter. Bags of leaves, boxes of wood, and trucks full of grass clippings, tree trimmings, and wood shavings are just a few examples. This vegetation does not have to be buried

This North Carolina hog farmer meets 75 percent of his farm's electricity needs by harvesting methane gas from the waste of his 4,000 hogs. Here, he steps on a pipe that transports the methane gas from a two-acre covered waste lagoon to the generator that powers the electricity.

forever. Rather, it can be used for making bioenergy. This would save space in the landfills, and the world would be a much cleaner place.

Like manure lagoons, landfills are also a source of methane gas. The methane gas becomes trapped beneath landfills as material piles up. This gas, if "tapped" from landfills with pipes and other technologies, can be used to make energy.

Energy Crops

The last major supply of the world's biomass fuel is from so-called energy crops. Energy crops, unlike normal crops that farmers

This display of energy crops at the Beacon biomass energy conversion facility in Nevada, Iowa, includes ground corn stover, obsolete seed corn, ground switchgrass, fescue grass, and wood chips. Researches are currently trying to identify the best plant sources for biofuel.

grow for food or fiber (cotton, for instance, is a type of fiber crop), are produced specifically for making bioenergy. Energy crops don't take the place of food crops. Instead, they are planted on land that is not needed for growing food. There is lots of land like this in the United States. In fact, the U.S. Department of Energy estimates that nearly 100 million acres of land could be used to produce energy crops in this country alone. If you think of all the land available throughout the world, the potential for energy crops as a source of biomass is staggering. Energy crops range from plants like switchgrass and alfalfa, which can also be used as hay for feeding animals, to fast-growing trees like willows, poplars, and eucalyptus. Scientists are currently researching new energy crops hoping to find plant species that produce the most energy in the least amount of time.

GALLONS
5
4
Systems
CFN

2 HISTORY AND DEVELOPMENT

People have burned logs, straw, and animal waste throughout history. Today, in many developing third-world countries, where energy is often scarce, villagers set rows of cow patties out in the sun to dry so they can be used as fuel for cooking or for heating homes. It's a process these people have relied on for generations. And because they need very little energy to live, they will likely continue to burn discs of dried manure and other forms of biomass for years to come.

In other parts of the world, the use of biomass for fuel is no longer widespread. The industrial age, which began in the 1800s, introduced coal and oil as cheap and convenient alternatives to biomass. Developed nations have, ever since, looked to fossil fuels

FOSSIL FUELS AND AIR POLLUTION

According to the U.S. Environmental Protection Agency, fossil fuels burned to run cars and trucks, heat homes and businesses, and power factories are responsible for:

- 98 percent of U.S. carbon dioxide emissions
- 24 percent of methane emissions
- 18 percent of nitrous oxide emissions

for power. Today, most of the industrialized world gets its energy from fossil fuels. The United States, the world's largest energy consumer, gets more than 80 percent of its energy from coal, oil, and natural gas.

Relying on fossil fuels has come at a price. According to the United States Environmental Protection Agency (EPA), since the industrial revolution, there have been huge increases of airborne pollutants. Concentrations of carbon dioxide released into the air by the burning of fossil fuels have increased nearly 30 percent since the late 1800s. Methane concentrations have more than doubled. Nitrous oxide, another pollutant, has seen a 15 percent increase. These pollutants are a prime reason for the thick haze called smog seen over many cities. People can develop asthma and other respiratory illnesses from breathing smog.

As if smog wasn't enough, scientists have shown that the unnatural increases in carbon dioxide levels are the major culprits in global warming. Global warming occurs when the earth warms up

The smog that hangs over many American cities, such as Los Angeles, is a direct result of the burning of fossil fuels. Environmentalists encourage the development of alternative energy sources such as biofuels to reduce the level of carbon dioxide released into the atmosphere.

as a result of the heat-trapping gases in the atmosphere. EPA records show that the average temperature at the surface of the earth has increased by about 1 degree Fahrenheit since the late nineteenth century. This may not sound like much to you, but to the earth's environment, it is. In addition, the ten warmest years of the twentieth century have all occurred in the last twenty years.

Other signs of global warming include decreased snow in the Northern Hemisphere, less ice floating in the Arctic Ocean, and a rise in sea levels throughout the world. All of these changes can have serious consequences for plant life, wildlife, ecosystems, and people!

FACT

American drivers fill their vehicles with nearly 100 billion gallons of gasoline every year.

Global warming has forced governments around the world to rethink their energy strategies. Many are looking for ways to limit the amount of greenhouse gases released into the atmosphere. To do this, people will have to decrease the amount of fossil fuels they use.

Today, biomass is becoming more and more popular as a source of energy, but it's still a very small part of the world's overall energy picture. In the United States, for example, biomass is used for just 3 percent of all heating and electric power needs. Generating electricity causes the largest amount of industrial pollution in the country. We have a long way to go.

Ethanol and Biodiesel

The story of ethanol and biodiesel is a story of economics. Despite the environmental benefits of biofuels, it has been almost impossible for biofuels to compete with cheaper fuels like gasoline and petroleum-based diesel.

This history begins in the early 1900s, when Henry Ford designed the Model T, which ran on a blend of ethanol and gasoline. Fuel stations across the country sold a gasoline-ethanol mix.

But Ford and his colleagues ran into trouble when the price of corn climbed too high to keep this blend inexpensive. When the price of corn was high, so was the price of ethanol. When fuel makers began to manufacture pure gasoline, farmers took a big financial blow.

Ford Model T's line a showroom in 1925. The idea of using corn-based ethanol to power cars dates back to the beginning of the history of the automobile.

Years later, in the late 1920s and 1930s, there were efforts to bring back ethanol for sale on a large scale. At one point, more than 2,000 Midwest service stations sold an ethanol-gasoline blend they called "gasohol." Produced at a single fermentation plant in Kansas, it appeared that gasohol might become the fuel of choice.

But gasohol met the same fate as the earlier ethanol fuels. People weren't willing to pay to have corn converted into fuel. In the 1940s, the Kansas ethanol plant shut down; petroleum, the cheaper alternative, took over.

In 1990, the Clean Air Act required, among other things, that oxygenated fuels be sold in parts of the country where pollutants like carbon monoxide had reached dangerous levels. Oxygenated fuels reduce the amount of pollution emitted by vehicles.

Dr. Steven Paul, the inventor of a fuel made partially from organic waste, poses with a flexible-fuel Ford Taurus that can run on both regular gasoline and the biofuel he invented.

When ethanol is blended with gasoline, it dramatically reduces the amount of toxic chemicals released into the atmosphere when that gasoline is burned. There are different kinds of ethanol-gasoline blends. E10, a gasoline-ethanol, consists of 10 percent ethanol. This has been the most common fuel available at gas pumps. E85 and E95, on the other hand, contain 85 and 95 percent ethanol respectively. These fuels, which burn even cleaner than E10, are not yet available across the country. They are primarily used in some government vehicles, "flexible-fuel" passenger vehicles, and urban-transit buses.

One of the reasons E85 and E95 are not as popular as E10 is that most vehicles today are not built to run on ethanol. E10, however, can

FACT

Vehicle emissions cause nearly 60 percent of urban air pollution. The use of biofuels like ethanol and biodiesel can reduce these emissions significantly—by as much as 90 percent when it comes to carbon dioxide, carbon monoxide, and smog.

be used in any gas-powered automobile. Next time you're at a gas station, take a look at the writing on the pump to see if gas with ethanol is available.

The history of biodiesel is similar to that of ethanol. Rudolf Diesel invented the diesel engine more than a century ago. The original engine ran on peanut oil, but Diesel's goal was for his engines to operate on a variety of vegetable oils.

Unfortunately, the vegetable oils that Diesel hoped would power his engines were too expensive to compete with cheap petroleum-based diesel fuel. Petroleum-based diesel was easy to make, quite efficient, and plentiful. For consumers, it became the fuel of choice.

Today, with scientific proof that global warming is real, some people are working to bring back biodiesel as an alternative to petroleum diesel, a heavy polluter. Many buses now operate on biodiesel, as do some large trucks, airport shuttles, boats, military vehicles, and other government vehicles.

Stop Global
Energy
Cents
Power

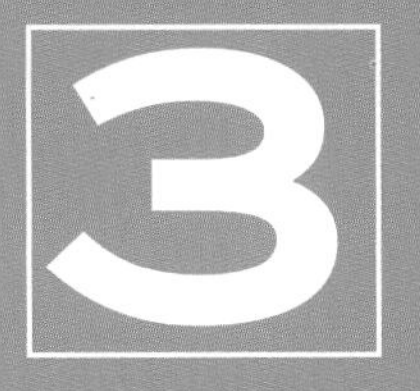

3 THE POLITICS OF ENERGY

Reading about biomass and other clean and renewable energy resources like wind and solar power, you might wonder why fossil fuels are used at all. Why would we burn coal and oil and send countless tons of pollutants into the sky when we could burn biomass instead and prevent the further buildup of greenhouse gases?

Well, in an ideal world, we wouldn't. But thanks to politics and economics, the picture is not so clear.

Some people say that big businesses—like oil companies, for instance—have serious influence in the political world. They contribute money to political campaigns. They work to place pro-business politicians in key positions in the federal and state

President George Walker Bush, pictured here in a meeting with energy secretary Spencer Abraham, has been criticized by environmental groups for not being seriously committed to exploring alternative energy sources.

governments. The companies help the officials get elected, and once that is achieved, they expect politicians to return the favors.

Many observers point at the current U.S. administration's energy policy. They feel that it rejects conservation and fails to put significant emphasis on the use of alternative energy resources because current leaders have ties to the oil industry. As a result, many people believe that the U.S. government continues to push policies that make the United States heavily dependent on oil and coal. If this is so, the United States will continue to import oil from foreign countries, despite pressure from the public to make "alternative energy" a priority.

Importing oil from foreign countries comes at a price. Right now, the United States depends on the Organization of the Petroleum Exporting Countries (OPEC) for more than 50 percent of its oil supply. Unfortunately, many of the countries that belong to OPEC have

WANTED: MORE BIOFUEL

There's no doubt about it: Ethanol and biodiesel production is on the rise. According to the U.S. Department of Agriculture (USDA), in 2000, ethanol producers churned out 246 million gallons more than they did in 1999. Also in 2000, there was a 36-million-gallon increase in production by those who make biodiesel.

This leap in production spells good news for both the environment and for American farmers who grow soybeans, corn, and other potential biofuels. Farmers can make more money if ethanol and biodiesel producers are buying their crops for biofuel. The environment also benefits. Increased production of ethanol and biodiesel means less dependence on pollutant-heavy traditional energy sources like coal, oil, and natural gas.

In the coming years, production of ethanol and biodiesel is expected to increase even more. One reason is the USDA's Bioenergy Program, which aims to encourage industry to use agricultural products like corn and soybeans. By promoting the use of these products in the creation of bioenergy, the USDA hopes to meet this goal by paying millions of dollars to commercial ethanol and biodiesel producers when they increase their bioenergy production by a set amount.

In a USDA press release dated August 30, 2001, USDA chief economist Keith Collins explains that the program "expands demand for corn and other grains used in ethanol production and creates new markets for oilseed crops. The program means increased net returns for ethanol and biodiesel processors, which will encourage expanded production capacity for these fuels." In other words, the more we use biofuels, the more biofuel crops we will need. And as the demand for biofuel crops increases, so will the market for farmers.

A herd of musk oxen graze in the part of the Arctic National Wildlife Refuge known as the 1002 area, which President George Walker Bush hopes to open to oil drilling. The president's plan faces strong opposition in Congress and among many environmental groups.

terrible human-rights records and are known sponsors of terrorism. In an age when we've declared war on terrorism, many Americans feel that this presents a difficult double standard. They tell us that, on one hand, the United States government says "no" to terrorism. But on the other hand, it politely asks those countries that support terrorism to sell us their oil (and keep the price down). Observers say that when the United States buys oil from these countries, money from the sale is used to fund the terrorists. They emphasize that the United States financially supports countries like Saudi Arabia because of their continued supply of oil.

BIODIESEL GETS THE THUMBS UP

On June 22, 2000, the U.S. Congress announced that biodiesel met the requirements imposed by the Clean Air Act Amendments of 1990. These laws set air-quality standards for the nation. They were created primarily to reduce pollution in urban areas. Tests on biodiesel showed that the fuel is not toxic to humans or the environment and does not contain the dangerous chemicals found in regular diesel.

As a proposed solution to this problem, the U.S. administration has recommended picking up the pace of oil drilling on American soil—places like the Arctic National Wildlife Refuge in Alaska. The idea is that if we increase the amount of oil produced in the United States, we can decrease the amount we need to import.

Unfortunately, this creates an entirely new dilemma. If we start drilling in pristine natural areas, we'll end up destroying more wildlife and more of our environment. If we replace some of the oil we import from foreign countries with oil from the United States, we would minimize our problems overseas. But in drilling for oil, we would do even more damage to our air quality, water quality, and overall environmental health.

Many people suggest that a better solution would be to consider the real needs of the human population: clean air, clean water, and good health. But these things aren't easy to achieve when politics and the financial needs of corporations get in the way.

4 THE PROS AND CONS OF BIOENERGY

As is the case with almost any form of energy in use today, bioenergy has its pros and cons. If you look at it closely, you can find both great things about it and things that aren't so good. When you consider, for instance, its potential for cleaning up the environment, bioenergy is one of the best energy options available. Likewise, its promise as a money-maker for farmers is huge.

On the other hand, the technology required to harness bioenergy from biomass has yet to be perfected, so it's still relatively expensive and difficult to buy.

Consider the benefits and drawbacks of bioenergy, then decide for yourself. Does bioenergy have what it takes to be the energy of the future?

BIOFUELS IN EUROPE

Americans aren't the only ones who recognize the promise of biofuels. This "fuel of the future" has taken Europe by storm as well. According to the journal *Europe Energy*, biodiesel production in Europe rose from 55,000 metric tons (121,253,000 pounds) in 1992 to 700,000 metric tons (1,543,220,000 pounds) in 2000. France led the way in terms of production, with an astounding 724,431,560 pounds of biodiesel in 2000.

The Benefits of Bioenergy

It's hard to overestimate the promise of bioenergy. Especially when it comes to the environment and the economy, bioenergy could turn the world on its head.

Environmental Benefits

Over the last twenty years, scientists have gathered convincing evidence showing Earth's climate is changing. In some ways, climate change is natural. Over hundreds and thousands of years, parts of the world undergo minor temperature swings. These temperature shifts affect everything from the amount of snowfall and rainfall to the ability of plants and animals to survive.

But in recent years, humans have come to rely more and more on the burning of fossil fuels for energy. As the population increases, more fossil fuels are used. As a result, Earth's normal climate has been disrupted. Records show that the climate has been heating up faster than normal. Scientists call this change global warming.

When we burn biomass for energy, carbon dioxide is also released. But if we replant an equivalent amount of biomass to replace what we use (something we can't do in the case of fossil fuels), the result is that no more carbon dioxide is released during burning than is consumed during photosynthesis. This self-contained loop is known as a closed carbon cycle. Carbon is taken in and released at the same rate, keeping the environment stable. Unfortunately, the cycle is broken when we burn fossil fuel.

One of the major reasons for global warming is the burning of fossil fuels, like coal, oil, and natural gas. Fossil fuels are the ancient remains of biomass, which contains much of the energy it did when it was still alive, plus chemicals and minerals picked up from the earth over the ages.

When we burn fossil fuels for energy, the carbon dioxide that was removed from the ancient atmosphere millions of years ago during photosynthesis is returned to our present-day atmosphere. Scientists refer to carbon dioxide as a greenhouse gas. As greenhouse gases accumulate, they trap heat from the Sun close to the surface of the earth, which results in global warming and air pollution.

When fossil fuels are burned, sulfur and nitrogen oxides are released into the atmosphere. These chemicals are the major components of acid rain, which can severely injure or destroy plants and wildlife. Biomass hasn't spent millions of years undergoing transformation beneath the ground. It hasn't picked up chemicals from the earth, and it does not contain these dangerous chemicals. So when biomass is burned, it doesn't lead to the creation of acid rain.

Operators of many coal-fired power plants are reducing their sulfur emissions through a process called co-firing, in which biomass is burned with the coal. By reducing the amount of coal needed to produce an equivalent amount of energy, the amount of sulfur released into the atmosphere is also reduced.

Other environmental benefits of using biomass for energy include reduced soil erosion and water pollution. Fewer wildlife habitats are destroyed. Biomass crops can stabilize soil that might otherwise wash away during heavy rains or flooding. The plant roots hold the soil, preventing it from being swept into streams and lakes.

If crops are planted close enough to the water, they can provide shade for aquatic animals like fish. Some aquatic species require cooler water temperatures for their survival, especially during the hotter months of the year. Other animals, like certain species of birds, make seasonal homes among biomass crops, using them for shelter from the weather.

Finally, there's the issue of landfill space. Millions of tons of agricultural waste end up in landfills every year. As more biomass crops and remains from the agricultural industry are converted to bioenergy instead of buried in landfills, the more landfill space will be saved.

Economic Benefits

Biomass is a great source of energy for economic reasons, too. The U.S. Department of Agriculture estimates that 17,000 jobs are created with the production of every one million gallons of ethanol. In addition, the Electric Power Research Institute has shown that producing five quadrillion Btus (British thermal units) of electricity on fifty million acres of land can increase overall farm income by $12 billion annually. When you consider that the United States alone consumes about ninety quadrillion Btus annually, this is a significant amount of money.

Another plus on the economic side is the promise bioenergy holds for reducing the United States's dependence on foreign nations for oil. When we increase our use of bioenergy, we decrease our need for oil.

According to the American Biomass Association, an organization that promotes the use of biomass in the United States, Americans could reduce the amount of oil imported by 50 percent by turning to biomass for part of their energy needs. Americans

Rice farmers, such as Ken Collins in Butte County, California, are working with government officials to build a plant that will refine rice straw into ethanol.

currently spend about $50 billion on imported oil every year. If half that money went to local farmers instead of foreign countries, it would help the U.S. economy tremendously.

By decreasing dependence on foreign oil, the United States could increase its national security. It is expensive to defend access to foreign oil. The United States, for example, stations troops in places like Saudi Arabia to make sure it can always get the oil it needs. When there is conflict or war in an oil-providing country, the United States finds that it has limited access to that oil. Using biomass for energy would eliminate much of this problem.

Growing biomass for energy helps farmers survive. Farmers can use energy crops to help stabilize their incomes. When a farmer produces just one or two products, like beef and milk, for instance, there is the risk of losing money. If demand for beef and milk decreases, the prices go down. However, if the farmer has a large energy crop on the same farm, there may be a cushion from these kinds of financial problems. When the beef price goes down, the farmer can always pay the bills with the proceeds from the harvest of his or her energy crop.

Farmers stand to make more money by growing biomass crops, and they may also save time and labor. Biomass crops are usually easy to grow. They require very little tilling and few fertilizers once they're in the ground. Quite often, biomass crops need to be planted only once. They come up naturally, year after year.

It has been shown that using biomass to meet a nation's energy requirements leads to valuable boosts for rural economies. Because biomass is bulky and difficult to transport, the conversion facilities used to turn it into fuel are often located in rural areas, near the farms where the biomass is grown. This means more jobs for people living nearby, as the facilities must have staffs to keep them running.

Using ethanol and biodiesel fuel will decrease auto emissions, the pollutants produced during normal operation of a vehicle. When motors burn gasoline or diesel fuel, large amounts of toxic chemicals are emitted from the exhaust. With the use of ethanol or biodiesel, these pollutants are significantly decreased.

THE PROBLEMS WITH BIOENERGY

The good news is there aren't many drawbacks to using biomass for energy. The bad news is that these drawbacks are substantial.

Perhaps the biggest problem is price. At the moment, bioenergy is more expensive than energy produced through the burning of fossil fuels. Fossil fuels are cheap. And since most consumers choose to spend as little money as possible, no matter what the ultimate cost to the environment, fossil fuels remain the fuel of choice in many parts of the world.

For bioenergy to become more affordable, we must produce more of it. But it's difficult to get providers to make more if there isn't a demand. Bioenergy would be cheaper if more people bought it. But people won't buy it because it's too expensive.

Another challenge for bioenergy is that its technology needs to be more affordable and more efficient. However, it shouldn't be long before bioenergy technology is equal to, or better than, that for fossil fuels.

Biomass from landfills has problems, too. Since much of the biomass in landfills is mixed in with nonorganic materials, removing the useful biomass from the massive piles of trash is time consuming and costly. Burning regular trash along with the biomass sends toxic chemicals into the atmosphere. For landfills to be useful, materials must be sorted as they arrive to assure that biomass is set apart from materials that can't be burned for energy.

Landfills contain inorganic material that must be sorted out before their biomasses can be converted to fuel. The process, which can be expensive, poses health risks to the sorters.

Last but not least, bioenergy may not be as great an option for the environment as solar or wind energy. Burning biomass in a sustainable way does not reduce the amount of carbon released into the atmosphere. Instead, it keeps levels constant. Solar and wind energy, on the other hand, have no emissions whatsoever. With solar and wind energy, atmospheric carbon levels ultimately decrease.

5 INTO THE FUTURE

The world today is very different from what it was 100 years ago. The global population hit six billion in the late 1990s, and is currently increasing by about 70 million people per year. There were just 2.5 billion people on Earth in 1950. Technology today—including computers and the Internet, cell phones and satellites, spaceships and supersonic aircraft—is much more advanced than it was in the early twentieth century. Today we have electric razors, electric coffeemakers, and electric blow-dryers. We drive to school. We drive to work. We drive to the store. And we drive just to drive.

These advances have, for the most part, allowed people to work with greater efficiency, travel far more easily, and communicate with each other like never before. But they've also increased our dependence on energy harnessed from the world's natural resources.

Walking requires a healthy body and a supply of food and water, but an average car needs oil and gasoline to operate. You can put a pen to paper and write a letter by hand, or you can turn on your computer and send an e-mail to your friend. But if you choose to write an e-mail, you'll get nowhere without electricity.

This tremendous and constantly growing demand on the world's energy resources has resulted in a global energy crisis. We're not about to run out of energy, but we need to use fuel that doesn't pollute the air or water and doesn't destroy the forests. Biomass and bioenergy may be the answer.

For bioenergy to compete with other forms of energy, however, many things must happen. First, we must establish the true cost of fossil fuels. People rarely consider the real costs of coal and oil, which goes beyond the price that we pay at the gas station. There are increased health costs for treatment of diseases caused by air pollution. It costs money to clean up oil spills and leaks from storage tanks. And it costs money to fight the effects of global warming and acid rain. When the real costs are added up, fossil fuels don't seem so cheap.

Second, we must continue to research new technologies. When it comes to bioenergy, there is a lot of work to be done. The U.S. Department of Energy is currently conducting research in partnership

Gary Tumbleson, of the National Corn Growers Association, uses a model version of full-cell technology to demonstrate how ethanol can power automobiles.

with universities, other government agencies, and private companies who are involved in agriculture, forestry, and the environment. By pooling their resources, these diverse groups expect to make more progress than if they worked independently.

The focus of this research includes the following:

Wood Energy Crops

One goal is to develop better sources of wood energy. Researchers are looking for ways to increase wood biomass production by using the latest techniques in molecular genetics, breeding, and silviculture, a branch of forestry that deals with the development and care of forests. Researchers are also looking for that "perfect tree." The goal is to find wood sources that are fast-growing, require little care, and maximize energy production.

Plant Energy Crops

Another goal of current research is to find plants that can be produced economically in a range of climates and terrains. Scientists hope to develop plants that require few or no chemical pesticides and herbicides and little tending by farmers, and that can be used for energy products such as ethanol or biodiesel. One promising crop is switchgrass, which is highly productive, easy to maintain, and grows just about anywhere.

Environmental Sustainability

Researchers are working to prove that biomass crops can be established, managed, and harvested in a way that is good for the environment. By collecting data at research sites, scientists can study the effects certain crops have on water quality, soil health, and biodiversity.

Systems Engineering

It sounds technical, but it's really quite simple. The goal in systems-engineering research is to make biofuels and bioenergy more

A control room operator examines the control board at the Alliant Energy power plant in Chillicothe, Iowa, where switchgrass is being burned for energy.

affordable. This can happen if engineers design better ways to harvest, transport, and convert biomass into power.

Researchers are focusing on efficient ways to handle crop residues like corn stover. They're also collecting information on the moisture content of various plants, physical characteristics of certain grasses, and the best way to store harvested crops. They can use this data to design top-notch technology that will make the entire process—from planting seeds to producing energy at the power plant—more efficient.

Residue Research

The goal of the research by the USDA is to find new ways to turn residue from urban, industrial, agricultural, and logging and other forest wastes into energy. Researchers must consider environmental drawbacks and the benefits of certain residues. They must also determine the costs of specific biomass products, as well as reliable ways to handle and process it.

Ethanol and Biodiesel Research

Researchers are working to apply the latest scientific techniques and technologies to the biomass-to-biofuel conversion process. One goal is to find microorganisms that better ferment the sugars that are present in biomass. Another goal is to invent sensors that can reliably measure these sugars. This will help biofuel makers

CROP CITY

It's amazing how many different crops can be used in the production of ethanol or biodiesel. Here's a partial list.

- barley
- canola
- corn
- crambe
- flaxseed
- grain
- sorghum
- mustard seed
- oats
- rapeseed
- rice
- safflower
- sesame seed
- soybeans
- sunflower seed
- switchgrass
- wheat

decide what plants to use and the best time to harvest them. These and other improvements will lead to much lower ethanol and biodiesel prices.

Advances in technology make bioenergy a top contender to replace fossil fuels as the energy of the future. As energy consumers, we should understand our place in the process.

We want energy to power the machines that enable us to enjoy our lives. Each time we turn on a light, tune in a radio, play a video game, or watch television, we need power to keep those high-tech gadgets working. What's it worth to you to keep that energy flowing?

GLOSSARY

additive A substance added to another substance in order to make it work in a particular way.

biodiesel A type of liquid fuel made from biomass.

biodiversity Variety in an environment as seen in the number of different plant and animal species living there.

bioenergy Energy drawn from biomass.

biofuels Fuels made from biomass.

biomass Plant materials and animal waste that can be used as fuel.

carbohydrates Energy-rich compounds found in plants.

carbon A substance found in all living organisms and in their remains.

carbon dioxide A gas released into the atmosphere with the burning of fossil fuels and through the normal breathing process of animals. A limited amount of carbon dioxide is essential for life.

economics The science that deals with money and the economy.

emissions Substances discharged into the air.

energy The capacity for doing work.

energy crop Plants that are grown to be used as fuel.

environmentalist A person concerned about the health of the natural world.

ethanol An alcohol that can be used as fuel in automobiles.

fermentation A process by which carbohydrates in plants are converted into sugar and then into alcohol.

fossil fuel A fuel formed in the earth from the remains of plants or animals.

global warming The rise in temperature of the earth due to a combination of natural and unnatural causes.

greenhouse effect When the gases in the atmosphere, especially carbon dioxide and methane, trap heat close to the earth's surface.

methane A gas that that is produced when organic matter decomposes or when coal is burned.

natural resource A material used by humans that comes from nature.

nitrous oxide A polluting gas made of nitrogen and oxygen.

photon A small amount of energy from light.

photosynthesis A process by which plants convert energy from light into food.

pollutant Something that creates harmful waste.

renewable energy Energy that can replenish itself as it is used, so it never runs out.

stover crop "Leftovers," including things like corn leaves, stalks, and cobs, that can be used for making bioenergy.

toxic Poisonous.

FOR MORE INFORMATION

American Bioenergy Association
314 Massachusetts Avenue NE, Suite 200
Washington, DC 20002
(703) 516-4444
Web site: http://www.biomass.org

Canadian Renewable Fuels Association
31 Adelaide Street East
P.O. Box 398
Toronto, ON M5C 2J8
(416) 304-1324
Web site: http://www.greenfuels.org

CANMET Energy Technology Centre
1615 Lionel-Boulet Boulevard
Varennes, Quebec J3X 1S6
(450) 652-4621

Energy Efficiency and Renewable Energy Network (EREN)
U.S. Department of Energy
1000 Independence Avenue SW
Washington, DC 20585
(800) dial-DOE (342-5363)
Web site: http://www.eren.doe.gov/RE/bioenergy.html

National Agricultural Library
United States Department of Agriculture
10301 Baltimore Avenue
Beltsville, MD 20705
(301) 504-5755
Web site: http://www.nal.usda.gov/ttic/biofuels.htm

REPP-CREST
1612 K Street NW, Suite 202
Washington, DC 20006
(202) 293-2898
Web site: http://www.crest.org/bioenergy

Web Sites

Due to the changing nature of Internet links, the Rosen Publishing Group, Inc., has developed an online list of Web sites related to the subject of this book. This site is updated regularly. Please use this link to access the list:

http://www.rosenlinks.com/lfe/biof/

FOR FURTHER READING

Chandler, Gary. *Alternative Energy Sources* (Making a Better World). Breckenridge, CO: Twenty-First Century Books, 1996.

Graham, Ian S. *Geothermal and Bio-Energy* (Energy Forever). New York, NY: Raintree/Steck-Vaughn, 1999.

Houghton, Graham. *Bioenergy* (Alternative Energy Series). Riverdale, MD: Gareth Stevens Publishing, 1991.

BIBLIOGRAPHY

Brown, Paul. *Energy and Resources* (Living for the Future). New York: Franklin Watts, Inc., 1998.

Challoner, Jack. *Eyewitness: Energy.* New York: DK Publishing, 2000.

Chandler, Gary. *Alternative Energy Sources.* New York: Twenty-First Century Books, 1996.

Parker, Steve. *Earth's Resources* (Science Fact Files). Austin, TX: Raintree Steck-Vaughn, 2000.

Silverstein, Alvin. *Energy* (Science Concepts). Brookfield, CT: Twenty-First Century Books, 1998.

INDEX

A

acid rain, 42, 50
agricultural waste, 10, 16, 18–20, 43
air pollution, 13, 20, 26, 29, 31, 37, 42, 45–46, 50
American Biomass Association, 43
Arctic National Wildlife Refuge, 37

B

biodiesel, 13, 14–15, 28, 31, 35, 37, 40, 45, 46, 52, 54–55
bioenergy
- cost of, 46
- economic benefits of, 39, 43–46
- environmental benefits of, 39, 40–43
- problems with, 46–47

Bioenergy Program (USDA), 35
biofuels, 13, 17, 28, 31, 35, 40, 52–53, 54–55
- in Europe, 40

biomass
- definition of, 6, 10
- obtaining energy from, 10, 12–13, 39
- sources of, 15–23

C

carbon dioxide, 12, 26, 31, 41, 47
Clean Air Act, 29, 37
coal, 6, 25, 26, 33, 34, 35, 41, 42, 50
co-firing, 42
Collins, Keith, 35

D

Diesel, Rudolf, 31

E

Electric Power Research Institute, 43
energy, definition of, 5
energy/biomass crops, 21–23, 42, 43, 45, 52
ethanol, 13–14, 15, 28, 29, 30–31, 35, 40, 43, 45, 46, 52, 54–55
ethanol-gasoline blends, 29, 30

F

fermentation, 13–14, 29, 54
Ford, Henry, 28
forest industry, 16–18
fossil fuels, 6, 10, 25–26, 28, 33, 41, 42, 46, 50, 55

G

gasohol, 29
global warming, 26–28, 31, 41, 42, 50
greenhouse gases, 28, 33, 41–42

L

landfill waste, 16, 20–21
light energy, 11–12

M

manure/manure lagoons, 10, 20, 21, 25
methane gas, 20, 21, 26

N

natural gas, 26, 35, 41
nitrous oxide, 26

O

oil, 6, 25, 26, 33, 41, 50
  dependence on foreign nations for, 34–36, 37, 43–44
Organization of the Petroleum Exporting Countries (OPEC), 34–36

P

photons, 11, 12
photosynthesis, 11–12, 41

R

renewable energy, definition of, 6, 9

S

solar energy/power, 6, 9, 33, 47
stover, 20, 54
switchgrass, 23, 52

T

transesterification, 14

U

U.S. Department of Agriculture (USDA), 35, 43
U.S. Department of Energy, 14, 15, 16, 19, 22
  current research by, 50–55
U.S. Environmental Protection Agency (EPA), 26, 27

W

wind energy/power, 6, 9, 33, 47
windmills, 9

CREDITS

About the Author

Chris Hayhurst works as a writer from his home in Colorado.

Photo Credits

Cover, pp. 4, 7, 14, 15, 18, 19, 21, 22, 24, 30, 32, 34, 36, 38, 44, 47, 48, 51, 53 © AP/Wide World Photos; p. 8 © AFP/Corbis; p. 11 © Neil Rabinowitz/Corbis; p. 12 © Richard Hamilton Smith/Corbis; p. 27 © Kit Kittle/Corbis; p. 29 © Bettman/Corbis.

Editor

Jill Jarnow

Design and Layout

Thomas Forget